NORTH AMERICAN
FIELD GUIDES

REPTILES

Kristin J. Russo

Field Guides

An Imprint of Abdo Reference | abdobooks.com

CONTENTS

Tortoises

Turtles

Reptiles evolved from early amphibians 315 million years ago. Today, there are more than 10,000 different types of reptiles in the world. Several hundred species live in North America. Reptiles have scales and bony plates rather than fur or feathers. They are cold-blooded, meaning their bodies cannot produce heat. They rely on external heat sources, such as the sun, to help keep warm. Reptiles are vertebrates. This means they have a backbone. Many reptiles spend much of their time in water, but they do not breathe with gills as fish do. They have lungs and must pop their head above water for air. Some reptiles, such as turtles and tortoises, lay eggs on land. Others, such as most constrictor snakes, give birth to live young.

WHAT ARE REPTILES LIKE?

There are five main groups of reptiles: crocodilians, lizards, snakes, tortoises, and turtles. Some reptiles are tiny, such as the little brown skink lizard that is only 3 to 5.5 inches (7.6 to 14 cm) long. Others are large, such as the American alligator, which can grow to more than 11 feet (3.4 m) long and can weigh 1,000 pounds (453.6 kg). Reptiles live on land, underground, in the water, and in trees.

Some reptiles are venomous. Others are poisonous. Venomous reptiles release toxins into their victims through fangs when they bite. Poisonous reptiles do not need to bite. Humans and animals are exposed to toxins when they touch or eat poisonous reptiles.

Crocodiles, alligators, and caimans are included in the crocodilian group. The differences between crocodilians and other reptiles are inside their bodies. For example, a crocodilian heart is more complex. It has four chambers for pumping blood rather than three. Crocodilians also have a gizzard in their digestive system to help them break down tough foods. No other reptile has a gizzard—but birds do! In fact, crocodilians are more closely related to birds than they are to other reptiles.

Lizards can do something that no other reptile can do. They can detach, or drop, their tail. This defensive mechanism is called autotomy. It helps them escape predators. A lizard's tail has weak areas called fracture planes. When a predator bites the tail, the tail will detach at one of the fracture planes, depending on how much of the tail the lizard must lose to escape. The lizard will lose part or all of its tail, but it will not cause a wound.

Snakes have no arms or legs. So how do they move? There are many ways a snake's body can move. Most snakes move their body in a series of S shapes. This movement causes friction between the snake's scales and the ground and propels the snake forward.

Both tortoises and turtles have a protective shell on their body. The top part of the shell is called a carapace. The bottom half that covers the belly is the plastron. Tortoises spend most of their time on land. Their four legs are thick and strong like an elephant's and can bear the weight of their heavy shells as they walk. Turtles have four legs, but theirs are more like flippers. They spend more time in the water and need to be good swimmers.

WHAT ROLE DO REPTILES PLAY?

Reptiles are important members of the ecosystem. As both predators and prey, they help create balance in food chains. Plant-eating reptiles help spread seeds. This creates a healthy environment for many other animals that rely on trees, plants, flowers, and shrubs for food and shelter.

HOW TO USE THIS BOOK

Tab shows the reptile category.

The reptile's common name appears here.

LIZARDS

EASTERN FENCE LIZARD
(SCELOPORUS UNDULATUS)

The paragraph gives information about the reptile.

The eastern fence lizard is gray, black, or brown. Its belly is lighter with a splash of black flecks. This lizard has pointy scales that overlap and look like small beads. have black, wavy crossbands on their backs, and have bright-blue markings on their undersides. During season, males flash the bright-blue spots by doing ups" to attract females. They also flash their blue bellies to warn other males to stay away. Eastern fence lizards are tree dwellers and scurry quickly and easily up and down tree bark.

Fun Facts give interesting information about the reptile.

FUN FACT
When pursued, the eastern fence lizard will keep scurrying around and around to the opposite side of the tree from the predator, much like a squirrel, making it difficult to catch.

How to Spot features give information about the reptile's size, range, habitat, and diet.

HOW TO SPOT

Size: 4 to 7.3 inches (10.2 to 18.5 cm)
Range: Eastern and central United States
Habitat: Woodlands, forests, and grasslands
Diet: Insects, spiders, and invertebrates

22

6

EASTERN GLASS LIZARD
(OPHISAURUS VENTRALIS)

The eastern glass lizard is legless and looks more like a slithery snake than a lizard. These lizards vary in color. They can be light brown, yellow, or green, with lighter bands near their head. Its tail makes up more than half the length of its body. When the lizard is in danger, its tail detaches easily, shattering like glass. This is how the lizard got its name. The tail will regrow completely. Females lay clutches in early summer and watch over their nests until hatchlings emerge. This behavior is unusual for lizards.

HOW TO SPOT

Size: 18 to 43 inches (45.7 to 109.2 cm)
Range: East Coast of the United States
Habitat: Wetlands and coastlines
Diet: Insects, spiders, rodents, and small reptiles

AUTOTOMY

Like most lizards, the eastern glass lizard can perform autotomy. This means it can drop its tail when caught by a predator. The detached tail flops around on the ground, distracting the lizard's attacker long enough for it to escape.

23

AMERICAN ALLIGATOR
(ALLIGATOR MISSISSIPPIENSIS)

The American alligator is the largest reptile in North America. Its color depends on its habitat, but it can be olive, brown, gray, or black, with a light-colored belly. Bony plates cover its body, giving it a rough, dinosaur-like appearance. American alligators have five toes on their front feet, but their back feet have only four. Their nostrils are on top of their snout, making it easier to breathe while the rest of their body is underwater. American alligators are aggressive hunters, and their jaws are so strong they can break open a turtle's hard shell.

HOW TO SPOT

Size: 8.2 to 11.2 feet (2.5 to 3.4 m)
Range: Southern and southeastern United States
Habitat: Freshwater rivers, swamps, marshes, and lakes
Diet: Fish, birds, turtles, snakes, mammals, amphibians, insects, snails, and worms

HOW ARE ALLIGATORS AND CROCODILES DIFFERENT?

Alligators and crocodiles look alike, but there are differences between them. Alligator snouts are round. Crocodile snouts are pointed, like a triangle. Alligators live in fresh water. Crocodiles can live in salt water because they have a special gland that helps them get rid of extra salt in their bodies. When an alligator's mouth is closed, its bottom teeth are hidden. When a crocodile's mouth is closed, the fourth tooth on the bottom is visible.

AMERICAN CROCODILE
(CROCODYLUS ACUTUS)

The American crocodile has rough, grayish-green scales. Its legs are short, and its tail is long and thick. Like all crocodiles, American crocodiles have a pointed, V-shaped snout. Their nostrils and ears are on top of their head, so they can breathe and hear while the rest of their body is under the water. Unlike most reptiles that are born ready to live on their own, baby American crocodiles need their mother. Females stay with their young for several months, protecting them and helping them find food.

HOW TO SPOT

Size: 8 to 20 feet (2.4 to 6.1 m)
Range: Southern Florida in the United States and Mexico
Habitat: Saltwater swamps and rivers
Diet: Fish, birds, mammals, turtles, crabs, snails, and frogs

COMMON CAIMAN
(CAIMAN CROCODILUS)

The common caiman is also called the spectacled caiman because a ridge between its eyes makes it look like it's wearing eyeglasses. Its body is olive green with yellow or brown crossbands. These crocodilians are nocturnal, which means they are active at night. Common caimans remain quiet and stay underwater during the day. They may even hibernate in summer if a drought occurs and it becomes too difficult to find food.

FUN FACT
Florida is the only state in the United States where both crocodiles and alligators live.

HOW TO SPOT

Size: 3.5 to 6 feet (1.1 to 1.8 m)
Range: The Everglades in the United States and Mexico
Habitat: Swamps, ponds, creeks, mangroves, and rivers
Diet: Fish, amphibians, reptiles, waterbirds, and mammals

MORELET'S CROCODILE
(CROCODYLUS MORELETII)

The Morelet's crocodile is small compared with other crocodiles. It is grayish brown with dark blotches and crossbands, and it has a broad, V-shaped snout. Young crocodiles are yellow and have black bands. With thick, powerful tails and webbed back feet, these crocodiles are strong swimmers. Rather than chew their prey, they bite off and swallow large chunks. Females lay large clutches of up to 45 eggs in mound nests that they often share with other females. Mothers guard their nests and protect their hatchlings. Fathers will also help and protect their young. This is unusual behavior for reptiles.

HOW TO SPOT

Size: 5 to 9 feet (1.5 to 2.7 m)
Range: Mexico
Habitat: Freshwater lakes, rivers, and ponds
Diet: Fish, reptiles, birds, and mammals

BAJA BLUE ROCK LIZARD
(PETROSAURUS THALASSINUS)

The Baja blue rock lizard is hard to find in nature because its flat, bluish-gray body looks like a rock. It blends in almost completely with its surroundings. Dark crossbands run down its back and neck, and its tail is thin and pointed at the tip. During mating season in the spring, its head and tail turn an even brighter blue. Baja blue rock lizards are diurnal and forage for food during the day. They bask on rocks in the warm sun, but when danger approaches, their flat shape helps them scramble into rock crevices and caves to hide.

HOW TO SPOT

Size: 7 to 10 inches (17.8 to 25.4 cm)

Range: Baja California peninsula in northern Mexico

Habitat: Rocky caves, mountains, and forests

Diet: Leaves, fruits, flowers, insects, and small lizards

BROWN ANOLE *(ANOLIS SAGREI)*

The brown anole has dull, brown-gray skin with white or yellow speckles. Males have a bright-red or orange throat fan, called a dewlap. Not all females have a dewlap, and if they do, it is much smaller. Brown anoles are active during the day. They bask in the sunshine and hunt for food. In warm weather, female brown anoles lay a clutch of eggs every one to two weeks. In cool weather, brown anoles snuggle into tree bark or rotten logs.

HOW TO SPOT

Size: 5 to 8.5 inches (12.7 to 21.6 cm)

Range: Southern United States and Mexico

Habitat: Trees, shrubs, vines, and fences

Diet: Insects, grubs, spiders, other lizards and their eggs, aquatic invertebrates, fish, and their own detached tails

ATTRACTING FEMALE ANOLES

Male brown anoles flare their brightly colored dewlap to attract females during mating season. They also do this to warn other male anoles to stay away from their territory.

BRUSH LIZARD
(UROSAURUS GRACIOSUS)

The brush lizard is also called the long-tailed brush lizard. Its tail is twice as long as the rest of its body. Gray with dark crossbars, it has a dark-brown line on its side. This line looks like the branches of the creosote bushes where it lives. This makes it easier to hide from predators. Brush lizards mate in spring, and females lay a clutch in early summer. Hatchlings emerge after about two months. Although brush lizards can move fast, even during the heat of the day, they wait on perches for their prey to appear rather than hunting them down.

HOW TO SPOT

Size: 2.3 to 4.5 inches (5.8 to 11.4 cm)

Range: Southwestern United States and northern Mexico

Habitat: Low deserts and sandy areas near creosote bushes

Diet: Insects and spiders

CHUCKWALLA *(SAUROMALUS ATER)*

The chuckwalla is a large lizard with a dark, flat body. Adult males have black heads. Their trunks are often black, red, orange, gray, or yellow. Females and young chuckwallas have gray or yellow bands. Females lay a clutch of about six eggs in summer, and hatchlings emerge in the fall. They live and hunt alone. When they are in danger, they scramble into rocky crevices and wedge themselves into their stony burrows. This makes it hard for predators, such as rattlesnakes, kestrels, and hawks, to catch them.

FUN FACT
The name chuckwalla comes from the word *tcaxxwal*, which is the name for this lizard in the language used by the Cahuilla, a Native American people.

HOW TO SPOT

Size: 20 inches (50.8 cm)

Range: Southwestern United States and northern Mexico

Habitat: Rocky deserts

Diet: Insects and desert plants, especially the leaves of the creosote bush

COLLARED LIZARD
(CROTAPHYTUS COLLARIS)

The collared lizard is green and has tan, brown, blue-green, or yellow scales. It has two black stripes around its neck that look like a collar. Females are not as colorful as males, but bright-red splotches appear on their skin when they are ready to lay eggs. The red marks disappear after the female lays her clutch. The collared lizard does not lose its tail easily, and if it does drop it, it will not grow back.

HOW TO SPOT

Size: 10 inches (25.4 cm)
Range: United States and Mexico
Habitat: Desert grasslands and rocky terrain
Diet: Grasshoppers, crickets, and other lizards

FUN FACT
Collared lizards can run upright on their hind legs when they're in a big hurry. Few lizards can do this.

COMMON IGUANA
(IGUANA IGUANA)

The common iguana is a large, green lizard with a row of spikes from its neck to its tail. It has a throat fan called a dewlap. When an iguana is in danger, its dewlap swells to make the animal appear larger and scarier to its predator. Males make their dewlaps bigger to attract females during mating season. This iguana is arboreal and lives mainly in trees. Females leave the trees only to lay a clutch of eggs in soft ground. Young iguanas are a much brighter green than adults.

HOW TO SPOT

Size: 4.9 to 6.5 feet (1.5 to 2 m)

Range: Southern United States, Hawaii, and Mexico

Habitat: Shrubs and trees

Diet: Bird eggs, carrion, and vegetation, especially of the fig tree

CURLY-TAILED LIZARD
(LEIOCEPHALUS CARINATUS)

The curly-tailed lizard, also known as the northern curly-tailed lizard, is light brown or tan with black markings. It has a rounded snout. As its name implies, the curly-tailed lizard can curl its tail. It does this to look bigger and more dangerous when predators approach. If attacked, the lizard can detach its tail, but it won't grow back to be quite the same. The new tail will be harder to move and curl. Curly-tailed lizards hunt by remaining motionless and waiting for insects to amble by.

HOW TO SPOT

Size: 10.2 inches (25.9 cm)

Range: Southeastern United States

Habitat: Rocky coastlines

Diet: Insects, flowers, and fruits

DESERT GRASSLAND WHIPTAIL
(ASPIDOSCELIS UNIPARENS)

The desert grassland whiptail is dark brown or green and has small, coarse scales. It is long and slim, and its tail is longer than the rest of its body. Six yellow lines run from its head to its tail, which is blue at the tip. Desert grassland whiptails are solitary, which means they live alone and not in groups. They are diurnal, foraging and basking during the day. At night, they sleep in their burrows.

HOW TO SPOT

Size: 2.7 to 5.1 inches (6.9 to 13 cm)
Range: Southwestern United States and Mexico
Habitat: Low valleys and grasslands
Diet: Insects

PARTHENOGENESIS

All desert grassland whiptail lizards are female. They do not need to mate with a male to reproduce. They reproduce by a process called parthenogenesis. This means that the mother's cells multiply in such a way that new lizards are formed. Babies are not exact clones of their mother, however, because the reproductive material that makes new baby lizards is separated and recombined during the parthenogenesis process. Parthenogenesis is common in insects and other invertebrates, but it is rare in vertebrates, such as reptiles.

DESERT HORNED LIZARD
(PHRYNOSOMA PLATYRHINOS)

The desert horned lizard, also called the "horny toad," is a medium-sized lizard and not a toad as its nickname suggests. It is tan with dark, wavy blotches on its neck and flat body. A row of scales appears on each side of its throat. Some of these lizards have pointed scales on their back. Females lay clutches in spring, and hatchlings emerge about two months later. Desert horned lizards make themselves look bigger by puffing up with air when upset or threatened. To rest or hide from predators, they bury themselves in the sand or scurry into abandoned burrows.

FUN FACT

Desert horned lizards have small, hollow spots beneath their eyes that they can fill with blood. When a predator approaches, the lizard squirts blood from its eyes to scare its attacker away.

HOW TO SPOT

Size: 3.7 inches (9.4 cm)
Range: United States and Mexico
Habitat: Deserts
Diet: Insects, worms, and plants

DESERT IGUANA
(DIPSOSAURUS DORSALIS)

The desert iguana is a medium-sized lizard with a rounded snout and a long tail. It lives in the desert, and its pale-gray or white-and-tan scales help it hide from predators in the hot, sandy environment. It can stand high temperatures up to 115 degrees Fahrenheit (46°C). It hibernates in winter. In spring, it emerges from hibernation ready to breed. The desert iguana eats its own poop. This helps it digest the tough plants in its diet. Also, by eating its food twice, it gets more nutrients. It may seem gross, but eating poop is good for iguanas.

HOW TO SPOT

Size: 16 inches (40.6 cm)

Range: Southern United States and northern Mexico

Habitat: Desert scrublands and rocky stream beds

Diet: Insects, carrion, fecal pellets, and leaves and flowers, especially of the creosote bush

EASTERN FENCE LIZARD
(SCELOPORUS UNDULATUS)

The eastern fence lizard is gray, black, or brown. Its belly is lighter with a splash of black flecks. This lizard has rough, pointy scales that overlap and look like small beads. Females have black, wavy crossbands on their backs, and males have bright-blue markings on their undersides. During mating season, males flash the bright-blue spots by doing "push-ups" to attract females. They also flash their blue bellies to warn other males to stay away. Eastern fence lizards are tree dwellers and scurry quickly and easily up and down tree bark.

FUN FACT

When pursued, the eastern fence lizard will keep scurrying around and around to the opposite side of the tree from the predator, much like a squirrel, making it difficult to catch.

HOW TO SPOT

Size: 4 to 7.3 inches (10.2 to 18.5 cm)
Range: Eastern and central United States
Habitat: Woodlands, forests, and grasslands
Diet: Insects, spiders, and invertebrates

EASTERN GLASS LIZARD
(OPHISAURUS VENTRALIS)

The eastern glass lizard is legless and looks more like a slithery snake than a lizard. These lizards vary in color. They can be light brown, yellow, or green, with lighter bands near their head. Its tail makes up more than half the length of its body. When the lizard is in danger, its tail detaches easily, shattering like glass. This is how the lizard got its name. The tail will regrow completely. Females lay clutches in early summer and watch over their nests until hatchlings emerge. This behavior is unusual for lizards.

HOW TO SPOT

Size: 18 to 43 inches (45.7 to 109.2 cm)
Range: East Coast of the United States
Habitat: Wetlands and coastlines
Diet: Insects, spiders, rodents, and small reptiles

AUTOTOMY

Like most lizards, the eastern glass lizard can perform autotomy. This means it can drop its tail when caught by a predator. The detached tail flops around on the ground, distracting the lizard's attacker long enough for it to escape.

FIVE-LINED SKINK
(PLESTIODON FASCIATUS)

Also known as the American five-lined skink, the five-lined skink gets its name from the five cream-colored lines that run down the center of its back. There are two lines on each side and one straight down the middle. This medium- to large-sized lizard has short legs and is gray, brown, or black. Young five-lined skinks have blue tails. Adult males have a reddish-orange spot on their heads. Females lay a clutch of eggs in summer in moist soil or rotten logs. Hatchlings emerge about two months later. Like many lizards, the five-lined skink will detach its tail to escape predators.

HOW TO SPOT

Size: 5 to 8.5 inches (12.7 to 21.6 cm)

Range: Eastern United States

Habitat: Woodlands

Diet: Insects, spiders, and invertebrates

FLAT-TAIL HORNED LIZARD
(PHRYNOSOMA MCALLII)

The flat-tail horned lizard's body, not its tail, is flat. It has pointed scales that look like horns on its jaw and the back of its neck. Its tan, orange, or reddish-brown color helps it blend in with the sandy soil where it lives. It has a black line down the center of its back. Dark blotches mark each side of the line. These lizards have long, thick tails. They mate in spring, and females lay clutches of about 3 to 10 eggs. Sometimes females will lay two clutches per year.

HOW TO SPOT

Size: 2.2 to 3.4 inches (5.6 to 8.6 cm) long from snout to vent, not including the tail

Range: Southwestern United States and northwestern Mexico

Habitat: Deserts

Diet: Ants, beetles, and other insects

GILA MONSTER
(HELODERMA SUSPECTUM)

The Gila monster is the largest lizard native to the United States. Its scales are round and bead-like. Its thick, stout, black body is covered in pink, yellow, and orange bands and blotches. Its head and neck are broad, and its wide feet have long, sharp claws. Females lay between 5 and 12 eggs per clutch. They do not stay to care for their nest or their young when they hatch. Gila monsters are thought to be lazy foragers and hunters. They eat prey that is nearby. If food isn't available, they don't bother to eat. They store fat in their tails and can go months without food.

HOW TO SPOT

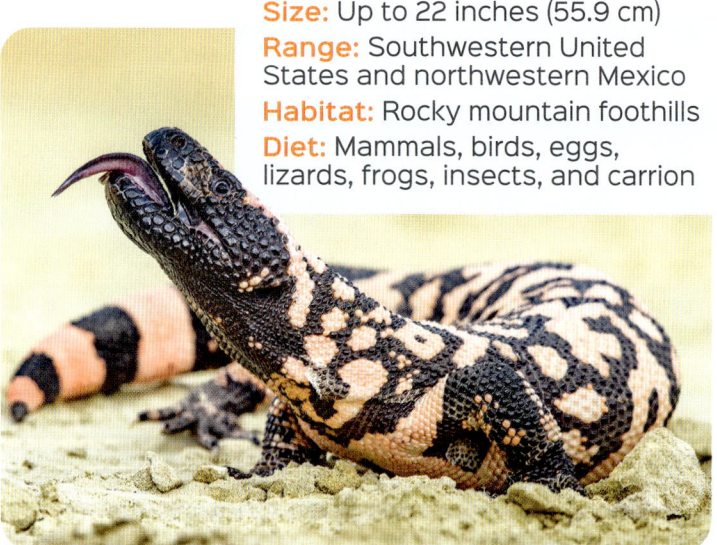

Size: Up to 22 inches (55.9 cm)
Range: Southwestern United States and northwestern Mexico
Habitat: Rocky mountain foothills
Diet: Mammals, birds, eggs, lizards, frogs, insects, and carrion

VENOMOUS LIZARDS

The Gila monster is one of very few venomous lizards in the world. The Mexican beaded lizard is another. These lizards deliver toxins through venom in their lower jaw. The venom is released from their gums as they chew their prey.

KNIGHT ANOLE *(ANOLIS EQUESTRIS)*

The knight anole is the largest species of anole. It is bright green with yellow stripes under its eyes and on its shoulders. Knight anoles are mostly arboreal. They are active during the day and tend to be alone. In normal conditions, they fiercely protect their territory from other anoles, but when it is cold, they huddle together for warmth. Breeding season takes place in summer. The male knight anole has a pinkish-white throat fan, or dewlap, which puffs up to impress females or scare off predators.

HOW TO SPOT

Size: 13 to 20 inches (33 to 50.8 cm)

Range: Southern tip of Florida in the United States and Cuba

Habitat: Mangroves, savannas, gardens, tropical forests, and trees along roadways

Diet: Fruits, insects, snails, small lizards, and sometimes frogs and nestling birds

SEED DISPERSAL

The knight anole performs an important role in the ecosystem as a seed disperser. Fruits are a regular part of its diet. Fruit seeds pass through its digestive system. They come out in its poop and can take root in the ground where they land.

LESSER EARLESS LIZARD
(HOLBROOKIA MACULATA)

The lesser earless lizard does not have any openings for ears on its head. Most lizards have outer ear holes, so this trait is unusual. This lizard spends much of its time burrowed in loose soil, and the lack of ear openings prevents soil and debris from getting under its scaly skin. Lesser earless lizards are small and gray with dark crossbands. Males have two black bars on the sides of their white bellies. The bars are surrounded by a blue color that gets brighter during mating season. Unless they are moving, they are difficult to spot.

HOW TO SPOT

Size: 2.5 to 3 inches (6.4 to 7.6 cm)
Range: Central and southwestern United States and northern Mexico
Habitat: Prairies and deserts
Diet: Insects

28

LITTLE BROWN SKINK
(SCINCELLA LATERALIS)

The little brown skink is a small, skinny lizard with a long tail and short legs. It is remarkable for the coppery-gold shimmer on its brown body. Its lighter belly is white or yellow. Dark stripes run down the sides of its body. Many skinks are arboreal, living in trees, but not little brown skinks. They almost never climb trees. Their body shape is best for living in moist, loose soil. Like many other lizards, the little brown skink will detach its tail to escape from predators, and its tail will grow back. Females lay several clutches throughout the summer.

HOW TO SPOT

Size: 3 to 5.5 inches (7.6 to 14 cm)
Range: Southern United States
Habitat: Abundant in many habitats, particularly woodlands
Diet: Insects, spiders, and invertebrates

LONG-NOSED LEOPARD LIZARD *(GAMBELIA WISLIZENII)*

The long-nosed leopard lizard is a gray- or cream-colored reptile with light and dark spots, like a leopard. It is considered a large reptile and has a long tail, sometimes twice the length of its body. Adults have gray throats. Females getting ready to lay eggs have orange or reddish markings on their sides. These lizards are diurnal. They hunt during the day by waiting under low shrubs for prey to amble by. The long-nosed leopard lizard is one of the few species that can run on its hind legs. If threatened, it will hiss and bite before running away on two feet.

HOW TO SPOT

Size: 3.3 to 5.8 inches (8.3 to 14.7 cm)
Range: Western United States and northern Mexico
Habitat: Deserts and semiarid areas with some vegetation
Diet: Insects, spiders, small snakes, mice, other leopard lizards, and plants

FUN FACT

The long-nosed leopard lizard has a healthy appetite, but it will sometimes bite off more than it can chew. These lizards choke to death if they try to eat prey that is too large to swallow.

MESQUITE LIZARD
(SCELOPORUS GRAMMICUS)

The mesquite lizard is a medium-sized reptile with a black, gray, or green, flat body. It has green-and-gray crossbars on its back, front legs, and tail. These lizards have long tails, often longer than the rest of their body. Males have a dark line on each shoulder and blotches on their necks that are blue and black. Males also have blue patches on their sides, but females do not have these markings. Mesquite lizards have sharp claws that help them climb trees and dig for insects.

HOW TO SPOT

Size: 3.9 to 6.9 inches (9.9 to 17.5 cm)
Range: Southern tip of Texas in the United States and northern to central Mexico
Habitat: Trees
Diet: Insects

FUN FACT

Male mesquite lizards compete with other males by having push-up contests. They try to attract females and show their strength by bobbing up and down as if they are doing push-ups.

MOJAVE BLACK-COLLARED LIZARD *(CROTAPHYTUS BICINCTORES)*

The Mojave black-collared lizard has a pair of black bands around its neck, making it look like it's wearing a collar. This lizard has a broad head, thick rear legs, and a chunky tail that is laterally compressed, or flat from side to side with the tallest section in the center. Males are shades of lighter brown to orange, and females are dark brown or black. Both have blotches and spots of white or brown, like the desert sand where they live. Mojave black-collared lizards are diurnal. They are active during the day, even in extreme heat.

HOW TO SPOT

Size: 8 to 14 inches (20.3 to 35.6 cm)

Range: Central, western, and southwestern United States

Habitat: Deserts

Diet: Insects, spiders, small lizards, small snakes, leaves, and flowers

MOLE SKINK *(PLESTIODON EGREGIUS)*

The mole skink is a small, slender lizard with a long tail and short legs. It is grayish brown and has two light stripes that run down the sides of its body. Its tail can be orange or red. In fact, it is the only lizard in its range that has a red tail. Mole skinks wriggle or "swim" through sand or loose soil, often disappearing in a flash as soon as they are discovered. Like other lizards, the mole skink will break off its tail to trick a potential predator.

HOW TO SPOT

Size: 3.5 to 6 inches (8.9 to 15.2 cm)

Range: Southeastern United States

Habitat: Scrublands, coastal dunes, and coastal islands

Diet: Insects, spiders, and invertebrates

NORTHERN ALLIGATOR LIZARD
(ELGARIA COERULEA)

The northern alligator lizard looks like a miniature alligator. Its rounded snout and brown eyes give it an alligator-like face. Its brown, gray, or green body is covered in rectangle-shaped scales, and dark blotches outlined in white make it look like its much larger reptile cousin. Its yellowish or light-green belly has light-green or dark lines running down the sides. Northern alligator lizards may look like alligators, but they move like snakes. They wiggle and swing their limbs from side to side. They are also very good swimmers and dive into the water quickly to escape danger.

HOW TO SPOT

Size: 10 inches (25.4 cm)

Range: Northern United States and southwestern Canada

Habitat: Woodlands, forests, grasslands, and coastlines

Diet: Insects, invertebrates, lizards, birds, and eggs

FUN FACT
The northern alligator lizard tucks its feet under its belly when it basks so that a predator cannot tell what type of animal it is. Predators might mistake it for a stick or a snake.

ORANGE-THROATED WHIPTAIL
(ASPIDOSCELIS HYPERTHRUS)

The orange-throated whiptail gets its name from the orange blotch on its throat, running from the tip of its mouth to its chest. On females, the color grows brighter during mating season. Its body is black, dark brown, or grayish in color with yellow or whitish stripes down its back. These lizards have a grayish-blue or white belly. Their whip-like tails are bright blue, which fades to gray as they age. They are active in the mornings and evenings, but they burrow to keep cool during the hottest time of day.

HOW TO SPOT

Size: 6 to 10 inches (15.2 to 25.4 cm)

Range: Southwestern United States and Mexico

Habitat: Rocky hillsides, coastal areas, streamsides, and semiarid brushy areas

Diet: Spiders, insects, and small lizards

PRAIRIE SKINK
(PLESTIODON SEPTENTRIONALIS)

The prairie skink is tan to olive brown in color with a light-colored stripe down the back and one or two wide, dark stripes along the sides. When they are born, they have a bright-blue tail, but this fades as they age. Males have a reddish-orange blotch on their head during mating season, which occurs during May and June. Females lay clutches in shallow burrows and watch over their eggs until they hatch.

HOW TO SPOT

Size: 5 to nearly 9 inches (12.7 to 22.9 cm)

Range: Northern, southern, and central United States, and southern and central Canada

Habitat: Tallgrass prairies

Diet: Insects, spiders, snails, and smaller lizards

FUN FACT
The prairie skink hibernates for a long time—seven months per year—from September to April. It has special fat reserves in its body to keep it alive while remaining underground in its burrow.

ROUNDTAIL HORNED LIZARD
(PHRYNOSOMA MODESTUM)

The roundtail horned lizard resembles a flat, round rock. Depending on the sandy soil color where it lives, this lizard can be gray, light brown, or yellow. Its shape and color help it blend in with its rocky habitat. In addition to a small head and short limbs, the lizard has a crest of horns on its neck that makes it look like a tiny dinosaur. Sometimes it makes its body round to appear more rock-like to trick a predator and keep it from attacking.

FUN FACT
The roundtail horned lizard sits near harvester ant or honeypot ant colonies. Honeypot ants are the lizard's favorite food.

HOW TO SPOT

Size: 4.3 inches (10.9 cm)

Range: Southern United States and north-central Mexico

Habitat: Rocky, sandy, and semiarid habitats with little vegetation

Diet: Insects

RUIN LIZARD *(PODARCIS SICULUS)*

The ruin lizard has a long head, a slender body, muscular legs, and a tail twice the length of its body. It is green, yellowish, or light brown with a white, grayish, or greenish belly. These lizards live alone. They are diurnal, which means they are active during the day. Ruin lizards burrow deep underground to hibernate during winter. Sometimes called Italian wall lizards, they are native to Italy and other European countries. In the United States and Canada, they are considered an invasive species, causing harm to the environment.

HOW TO SPOT

Size: 3.5 inches (8.9 cm) long from snout to vent

Range: United States and Canada

Habitat: Forests, woodlands, scrublands, pastures, and urban areas

Diet: Insects, spiders, small mollusks, crustaceans, reptiles, mammals, and plants

38

SAGEBRUSH LIZARD
(SCELOPORUS GRACIOSUS)

The sagebrush lizard is a small lizard with gray, brown, tan, or dull-green scales. A light-gray or tan stripe runs down its back, and two lighter stripes appear on each side. Males have bright-blue patches on their bellies, but females do not have these markings. Sometimes they are called sagebrush swifts for their rapid movements. They move quickly and jump easily from rock to rock. Often, they will hide in bushes or under rocks or logs.

HOW TO SPOT

Size: 1.9 to 3.5 inches (4.8 to 8.9 cm)
Range: Western and southern United States and northwestern Mexico
Habitat: Mountains and shrublands
Diet: Ants, termites, grasshoppers, flies, spiders, and beetles

SLENDER GLASS LIZARD
(OPHISAURUS ATTENUATUS)

The slender glass lizard is a long, slim reptile with brown or yellowish scales and a dark stripe down its back. It has no legs and is often mistaken for a snake. Slender glass lizards are diurnal and hunt for insects and small reptiles during the day. When attacked, their tail will break off in several pieces, like shards of glass. This is where they get their name. The tail will grow back, but it will not look the same as it did. It won't have stripes or markings.

FUN FACT
The slender glass lizard has a rigid jaw and cannot open its mouth wide enough to eat large prey. It cannot eat anything that is larger than its head.

HOW TO SPOT

Size: 29.5 inches (74.9 cm) long from snout to vent

Range: Northern, southern, and central United States and southern Canada

Habitat: Sandy prairies and fields, savannas, and grasslands

Diet: Insects, spiders, and small reptiles

SOUTHERN ALLIGATOR LIZARD
(ELGARIA MULTICARINATA)

The southern alligator lizard is brown, gray, or yellowish with red blotches on its back. There are dark bands and white spots on its back, sides, and tail. Its eyes are light yellow with round pupils. Sometimes there are dark lines or dashes on its belly. This lizard is active at different times of day depending on the temperature. It is usually diurnal, but it sometimes hunts at dusk just before the sun goes down. In hot weather, it can be nocturnal and move about at night. It will hide under rocks or low shrubs and is inactive in cool weather.

HOW TO SPOT

Size: 12 inches (30.5 cm)

Range: Mexico and the northwestern and southwestern United States

Habitat: Grasslands, open forests, and urban areas

Diet: Invertebrates, small lizards and mammals, birds, and bird eggs

TEXAS BANDED GECKO
(COLEONYX BREVIS)

The Texas banded gecko is a pinkish-brown color with bands of brown and yellow. As it ages, its blotches become more mottled. This gecko has small, grainy scales. It has large eyes with vertical pupils, like a snake. Unlike a snake, it has moveable eyelids. Its thick tail grows to the same length as its body. Its toes are thin. This lizard is nocturnal and hunts at night. Like most reptiles, it sheds its outer skin as it grows, and it will eat the skin as it sloughs off.

HOW TO SPOT

Size: 4 to 4.8 inches (10.2 to 12.2 cm)

Range: Southern United States and northeastern Mexico

Habitat: Deserts and dry, rocky canyons

Diet: Spiders, crickets, moths, and termites

FUN FACT
The Texas banded gecko makes a squeaky noise when it is picked up and handled.

TREE LIZARD *(UROSAURUS ORNATUS)*

The tree lizard, also called the ornate tree lizard, is one of the most abundant types of lizard in North America. This species has many different color patterns depending on where it lives. Many tree lizards are gray or tan with black, dark brown, and reddish markings. Some males have a blue dot in the center of their orange throat fan. Other males have an orange throat fan. All males have bright-blue or blue-green spots on their bellies. These lizards usually live in trees. Tree lizards that live in areas where there is little vegetation thrive in rocky areas with large boulders.

HOW TO SPOT

Size: Up to 2.3 inches (5.8 cm) long from snout to vent

Range: Western and southwestern United States and northern Mexico

Habitat: Forests, rocky terrains, fences, and buildings

Diet: Insects, insect larvae, and plants

FUN FACT

A male with a blue dot on its dewlap is usually more aggressive and territorial. Males with solid orange throat fans are calmer and less aggressive.

43

WESTERN WHIPTAIL
(ASPIDOSCELIS TIGRIS)

The western whiptail is orange or rusty brown. White, black, and tan markings help it blend in with its sandy environment. Its triangle-shaped head has a pointed snout. Rectangular scales are arranged in eight rows along its back. It also has white-and-black spots along its sides. Western whiptails do not live in groups. They have a higher metabolism than many other lizards and need to eat more frequently. They forage and hunt during the day. Females lay clutches once or twice per year in loose soil. They do not care for their eggs or their young.

HOW TO SPOT

Size: Up to 4 inches (10.2 cm) long from snout to vent
Range: Northern and southwestern United States and northern Mexico
Habitat: Deserts, desert scrublands, and woodlands
Diet: Insects, eggs, and other reptiles

ZEBRA-TAILED LIZARD
(CALLISAURUS DRACONOIDES)

The zebra-tailed lizard is a pale and slender lizard with small, grainy scales and long legs. Light and dark blotches cover its gray or light-brown body, and a bright-pink or orange spot appears on its throat. As the name suggests, they have a black-and-white pattern of bars and crossbands on their tails that resembles a zebra. Zebra-tailed lizards are active during the day. They will stay out basking on rocks even during the hottest part of the day. To hide from predators and to rest, they burrow in loose soil.

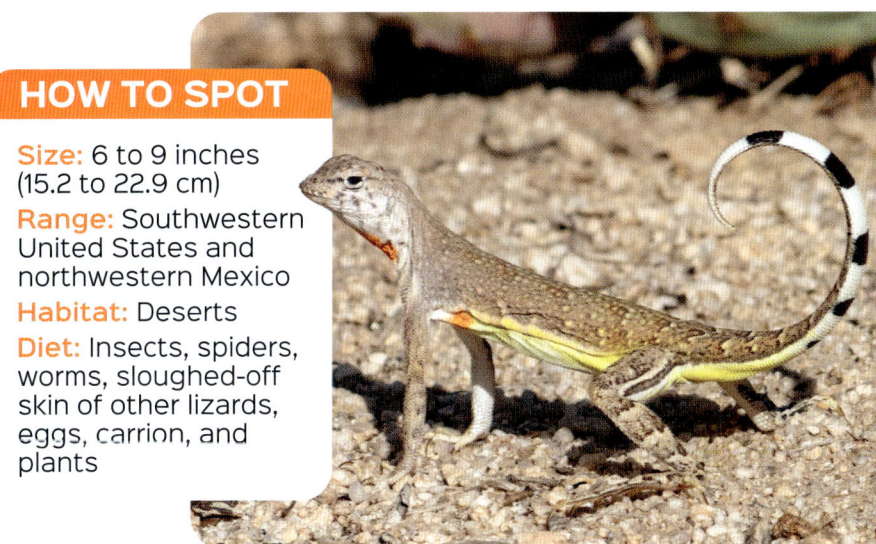

HOW TO SPOT

Size: 6 to 9 inches (15.2 to 22.9 cm)

Range: Southwestern United States and northwestern Mexico

Habitat: Deserts

Diet: Insects, spiders, worms, sloughed-off skin of other lizards, eggs, carrion, and plants

A WEDGE-SHAPED JAW

The zebra-tailed lizard has a countersunk jaw that is shaped like a wedge. The top half is slightly larger than the bottom half. When the lizard closes its mouth, the bottom half fits snugly inside the top half with no space in between. This keeps soil from getting into the lizard's mouth when it digs for food or to create a burrow. The lizard's unique jaw also helps it drink. It can take in rainwater from its rocky surroundings and the liquid will not dribble out.

ARIZONA RIDGENOSE RATTLESNAKE
(CROTALUS WILLARDI WILLARDI)

The Arizona ridgenose rattlesnake has a triangular head with scales ridged along its upturned nose, giving the snake its name. Its stocky body varies in color from a yellowish gray to a reddish brown. It has black markings from its head to its tail and two white stripes under its eyes. Young snakes have a gray, black, or yellow tail, but the color fades as the snakes age. They are diurnal and hunt during the day. They have retractable fangs that they can pull back into their jaw when they are not needed for biting.

HOW TO SPOT

Size: 11.8 to 26.3 inches (30 to 66.8 cm)

Range: Southwestern United States and northwestern Mexico

Habitat: Mountains, canyons, and forests

Diet: Lizards, mice, and centipedes

FUN FACT

The Arizona ridgenose rattlesnake is a pit viper. Pit vipers have two deep pits between their eyes and nostrils. The pits house special heat-sensing organs. When prey approaches, pit vipers can sense the rise in warmth in its environment. This gives them better aim when they strike. They deliver venom with their fangs.

ATLANTIC SALT MARSH SNAKE
(NERODIA CLARKII TAENIATA)

The Atlantic salt marsh snake is a light grayish tan with four dark, solid stripes that run from its head halfway down its back. From the midpoint of its back to its tail are dark bands or blotches. Its belly is black with yellowish spots. Atlantic salt marsh snakes are nocturnal and hunt for fish at night in shallow water. To avoid predators, they freeze and remain motionless so that they're harder to find. If this doesn't work, they will try to flee and will sometimes release a disgusting musky smell from glands at the base of their tail.

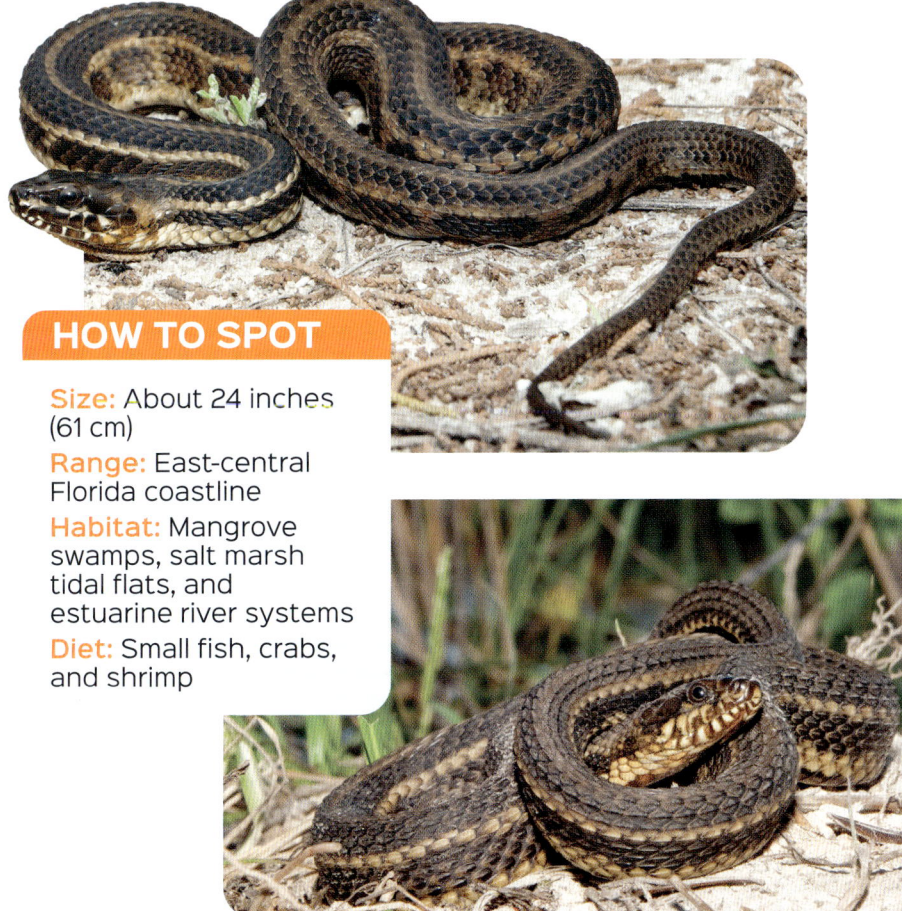

HOW TO SPOT

Size: About 24 inches (61 cm)

Range: East-central Florida coastline

Habitat: Mangrove swamps, salt marsh tidal flats, and estuarine river systems

Diet: Small fish, crabs, and shrimp

BAIRD'S RAT SNAKE
(PANTHEROPHIS BAIRDI)

The Baird's rat snake is yellow, orange, or pinkish with four dark stripes down the length of its body, which is small compared with other rat snakes. Its head is small, with little change in size between its face and neck. Its gray or yellow belly grows darker near the tail. These snakes are calm and not known to be aggressive. They are hard to observe in nature because they hide in dark spaces, such as rocky crevices and caves. Sometimes they shake their tail, pretending to be a rattlesnake, when they are threatened. Usually, they dart away to safety.

FUN FACT

Most reptiles are either nocturnal or diurnal, and scientists can predict whether they will be active at night or during the day. But Baird's rat snakes are cathemeral. Cathemeral animals sleep when they're tired and eat when they get the chance. Their behavior does not follow predictable patterns.

HOW TO SPOT

Size: 2 to 5 feet (0.6 to 1.5 m)

Range: Southwestern Texas in the United States and Mexico

Habitat: Semiarid and rocky areas

Diet: Rodents, birds, and lizards

BLACKNECK GARTER SNAKE
(THAMNOPHIS CYRTOPSIS)

The blackneck garter snake is slender and can be dark olive, brown, or gray in color. It has a black head and black blotches on its neck. A yellow or orange stripe runs down its back, and lighter stripes appear on its sides. Some blackneck garters have a dark checkered pattern between the stripes. This snake's saliva has a mild toxin that can produce an uncomfortable rash or reaction in humans. They mate in early spring, and females give birth to live young in the summer. They are known for moving fast and being excellent climbers.

FUN FACT

Most garter snakes live in communal dens. One colony of common garters, close cousins of the blackneck garter, was found to have about 8,000 snakes!

HOW TO SPOT

Size: 16 to 27 inches (40.6 to 68.6 cm)

Range: Southwestern United States, Mexico, and Guatemala

Habitat: Streams in valleys and canyons, woodlands, forests, desert scrublands, grasslands, tropical lowlands, and mountainous areas

Diet: Amphibians, fish, worms, and crustaceans

BLUE RACER
(COLUBER CONSTRICTOR FOXII)

The blue racer is a large, blue or gray snake. Its head is a darker color than its body, and it has a white chin and throat. Its belly is a lighter blue or white. Females lay 6 to 25 eggs in rotting wood or underground in June and July. Eggs hatch in late summer. Young blue racers are gray with blotches and spots that fade as they age. Like most racers, blue racers move quickly. They flee when startled, but if they can't get away, they act like rattlesnakes and shake their tail. Sometimes they strike, but they are nonvenomous, so their bite is fairly harmless.

HOW TO SPOT

Size: 3.9 to 5.8 feet (1.2 to 1.8 m)
Range: Canada and northern United States
Habitat: Open woodlands, meadows, prairies, marshes, and lake edges
Diet: Rodents, frogs, smaller snakes, birds, and insects

CALIFORNIA KINGSNAKE
(LAMPROPELTIS CALIFORNIAE)

The California kingsnake's scientific name means "shiny skin," and its scales do appear to be shiny and glossy. This snake varies in color, but most are black or dark brown with light-brown, yellow, or white bands. Its snout is light in color. These snakes hibernate in winter and emerge ready to mate in spring. They live in many different types of environments, including deserts, farmlands, woodlands, and marshes. Luckily, they are excellent at slithering, climbing, and swimming.

HOW TO SPOT

Size: 2.5 to 3.5 feet (0.8 to 1.1 m)

Range: Southern United States and northern Mexico

Habitat: Forests, woodlands, grasslands, marshes, farmlands, ranches, deserts, and brushy suburban areas

Diet: Small mammals, lizards, snakes, turtle eggs and hatchlings, amphibians, eggs, birds, and large invertebrates

RESISTANT TO RATTLESNAKES

The California kingsnake has a natural resistance to rattlesnake venom and is not harmed by it. A rattlesnake may strike and bite when a California kingsnake attacks, but it will do no good.

CAT-EYED SNAKE
(LEPTODEIRA SEPTENTRIONALIS)

The cat-eyed snake has large, bulging eyes with vertical pupils, just like a cat. Their eye structure helps these nocturnal snakes hunt at night. They are tan, buff, gold, or pale orange with dark blotches. A spearhead-shaped marking on their head points toward their tail. Cat-eyed snakes have a simple ear structure that allows them to detect vibrations in the ground, but they cannot hear sound. They deliver venom into their prey from fangs that sit at the back of their jaw.

HOW TO SPOT

Size: 18 to 24 inches (45.7 to 61 cm)

Range: North and Central America

Habitat: Grasslands, scrublands, and rain forests

Diet: Frogs, frog eggs, salamanders, and small reptiles

CHECKERED GARTER SNAKE
(THAMNOPHIS MARCIANUS MARCIANUS)

The checkered garter snake is tan, brown, or green with large, black markings in a checkered pattern from its neck to its tail. It has yellow stripes on its back and sides between the rows of checkers. Its head is small and not much wider than its neck. There is a dark blotch on each side of its head. Checkered garter snakes hibernate and emerge from hibernation in spring ready to breed. The snakes' saliva can cause a mild skin reaction, but their bite is not considered dangerous to humans.

HOW TO SPOT

Size: 18 to 24 inches (45.7 to 61 cm)

Range: United States and Mexico

Habitat: Grasslands, semiarid lands, streams, ponds, rivers, and irrigation ditches

Diet: Frogs, toads, small fish, lizards, small mammals, and invertebrates

HOW DID THE GARTER SNAKE GET ITS NAME?

Some scientists think the garter snake got its name from old-fashioned socks. The stripes on the snake look like the garters that people once used to hold up their socks. Other scientists think the name garter came from *garten*, which is the German word for "garden."

CORN SNAKE
(PANTHEROPHIS GUTTATUS)

The corn snake has a bright orange-brown body dotted with darker blotches outlined in black. Its belly is black and white and often has spots of orange. Because of its coppery color, this snake is sometimes mistaken for the venomous copperhead, but corn snakes are nonvenomous. They are constrictors and kill their prey by coiling around and suffocating it. These snakes are excellent climbers and are often found in trees searching for birds. On sunny days, they will crawl out of their hiding spots under logs and rocks to bask in the sun.

HOW TO SPOT

Size: 2 to 5.9 feet (0.6 to 1.8 m)
Range: Eastern and southeastern United States
Habitat: Fields, forests, and farms
Diet: Mice, rats, frogs, lizards, birds, and eggs

DIAMONDBACK WATER SNAKE
(NERODIA RHOMBIFER)

The diamondback water snake is brown or olive green with a series of diamond-shaped, black markings on its back. Its sides and belly are lighter brown or yellow. A series of thick, black bands connected to the diamonds also run along the sides of its body. Its belly has dark spots sometimes shaped like half-moons. Females give birth to 20 or more live young in late summer or early fall. Juveniles are lighter with a darker diamond pattern. To hunt for fish, this snake hangs from branches above the water. When danger approaches, the snake drops into the water and swims away.

HOW TO SPOT

Size: 4.9 feet (1.5 m)

Range: United States and Mexico

Habitat: Slow-moving bodies of water, streams, rivers, ponds, and swamps

Diet: Fish, frogs, and toads

EASTERN CORAL SNAKE
(MICRURUS FULVIUS)

The eastern coral snake has a long, slender, black body encircled by alternating red and yellow rings. The red rings are wide and the yellow rings are thin. It has a black head and a black-and-yellow tail. Eastern coral snakes are venomous, and most venomous snakes give birth to live young. This snake is different. Females lay clutches of eggs in the summer. When they are not hunting, these snakes rest and hide in sandy soil, leaves, logs, or abandoned gopher tortoise burrows.

HOW TO SPOT

Size: 18 to 30 inches (45.7 to 76.2 cm)

Range: Southern United States

Habitat: Forests and grasslands

Diet: Snakes, lizards, and rodents

FUN FACT

Eastern coral snakes look like the harmless scarlet snakes and scarlet kingsnakes. All three species have red, yellow, and black bands. Eastern coral snakes have red and yellow bands together. Scarlet snakes and scarlet kingsnakes have red and black bands together. Many people use the poem: "Red and black, friend of Jack; red and yellow, kill a fellow," to tell the snakes apart.

EASTERN HOGNOSE SNAKE
(HETERODON PLATIRHINOS)

The eastern hognose snake varies in color and can be red, green, orange, brown, gray, or black. Often it is blotched or checkered in black markings. Sometimes there are no dark blotches. Its belly is a lighter gray, yellow, or cream. As its name suggests, this snake has an upturned snout like a hog that it uses to dig for toads. Toads puff up with air to avoid being eaten, but eastern hognose snakes have a solution. Their sharp teeth pierce and deflate the air-filled toad, making it easier to swallow.

HOW TO SPOT

Size: 17 to 41 inches (43.2 to 104.1 cm)

Range: Central, eastern, and southeastern United States and southern Canada

Habitat: Forests, woodlands, prairies, and meadows

Diet: Reptiles, frogs, toads, small mammals, small birds, and fish

PLAYING DEAD

Eastern hognose snakes are dramatic when predators approach. They perform an elaborate fake death scene. They flip onto their back and shake as if having a seizure. Then they poop, excreting a foul-smelling musk. Finally, they lie motionless with their tongue hanging out. When the danger has passed, the snake will flip back over and slither away.

EASTERN MILK SNAKE
(LAMPROPELTIS TRIANGULUM TRIANGULUM)

The eastern milk snake ranges in color from pale gray to darker gray or brown. It has a pattern of reddish or brown blotches on its back. On the back of the head, they usually have a Y- or V-shaped mark that's gray or tan. Most milk snakes have a pattern of black-and-white checkers on their bellies. These snakes have the largest range of any snake in North America. Milk snakes evolved to look like more dangerous snakes, such as the venomous copperhead and coral snakes. This is called mimicry. Their appearance protects them from predators

HOW TO SPOT

Size: 24 to 36 inches (61 to 91.4 cm)

Range: Eastern and southeastern United States and southern Canada

Habitat: Forests, fields, rocky terrain, and areas near buildings and mammal burrows

Diet: Rodents, birds, and lizards

FUN FACT
The eastern milk snake's name came from a belief that the snakes lived in dairy barns to drink milk from cows. However, the milk snake goes into barns to hunt the rodents who live there, not to milk the cows!

EASTERN YELLOWBELLY RACER
(COLUBER CONSTRICTOR FLAVIVENTRIS)

The eastern yellowbelly racer is a slender, olive-green snake with a yellow belly. Its throat and neck are yellow, and its lips are yellow and white. No other markings appear on its head or back. Females lay clutches in late spring, and hatchlings emerge in late summer. Eastern yellowbelly racers move fast, especially when danger approaches. They are active and hunt during the day. These snakes "periscope" when they hunt by popping their head above the tall grass in search of their prey.

HOW TO SPOT

Size: 1.7 to 5.4 feet (0.5 to 1.7 m)

Range: Eastern and southeastern United States

Habitat: Open prairies, woodland edges, and grasslands

Diet: Rodents, frogs, smaller snakes, birds, and insects

FLORIDA COTTONMOUTH
(AGKISTRODON CONANTI)

The Florida cottonmouth has a thick, light-brown body with dark-brown crossbands that grow darker as the snake ages. Older cottonmouths are almost completely black. As its name suggests, the inside of its mouth is bright white. A dark brown or black stripe covers its eyes. Females give birth to live young in late summer or early fall. Young cottonmouths are much lighter in color than adults, and they have a yellow tail tip. Cottonmouths are venomous and deliver a dangerous bite. Young cottonmouths are born with fully functioning fangs.

FUN FACT

Though their venomous bites are dangerous to people and pets, Florida cottonmouths are not aggressive. Most bites occur by accident when the snakes are stepped on.

HOW TO SPOT

Size: 2.5 to 4 feet (0.8 to 1.2 m)

Range: Southeastern United States

Habitat: In or near any wet area and occasionally up to 1 mile (1.6 km) from water

Diet: Insects, small turtles, fish, mammals, birds, small reptiles, and carrion

FLORIDA WATER SNAKE
(NERODIA FASCIATA PICTIVENTRIS)

The Florida water snake is light brown or yellow with black, brown, or red bands across its body. Sometimes it has a black stripe down the center of its back. Its belly is yellow with red or black markings. Females give birth to live young in spring and summer. Young snakes are light in color and have red or black crossbands. They are active both at night and during the day, hunting in the water and basking on branches and logs. They are solitary and live alone. They defend themselves by giving off a foul-smelling musk when threatened.

HOW TO SPOT

Size: 2 to 4 feet (0.6 to 1.2 m)
Range: Southeastern United States
Habitat: Shallow fresh water, including swamps, marshes, ponds, lakes, streams, and rivers
Diet: Fish, frogs, and invertebrates

GLOSSY SNAKE *(ARIZONA ELEGANS)*

As its name suggests, the glossy snake has smooth and shiny scales. It varies in color, ranging from tan to brown or gray with spotted patterns on its back and a dark stripe beneath each eye. Its belly is white with no markings. Glossy snakes have a narrow, pointed head and a wedge-shaped, countersunk jaw. The bottom half of the jaw fits snugly inside the top half when the snake's mouth is closed, helping keep sand and soil from getting inside.

FUN FACT

Glossy snakes play an important role in the ecosystem. By burrowing under the earth, they aerate the soil. Aerated soil is looser, making it easier for plants to grow.

HOW TO SPOT

Size: 2.2 to 5.8 feet (0.7 to 1.8 m)
Range: Southwestern United States and northern Mexico
Habitat: Coastal scrublands, desert shrublands, grasslands, rocky shores, and deserts
Diet: Small mammals and lizards

GRAY-BANDED KINGSNAKE
(LAMPROPELTIS ALTERNA)

The gray-banded kingsnake has two common color patterns. "Blairi" snakes, named for naturalist William Franklin Blair, are gray with wide red or orange bands. "Alterna" are gray with thin red or orange bands. In both types, the bands are outlined in black. This snake has a wide head and bulging eyes with round pupils. Females lay clutches in early summer, and hatchlings emerge after about nine weeks. Not much else is known about these nocturnal snakes. They are difficult to observe in nature because they live in hot, harsh, mountainous environments that humans tend to avoid.

HOW TO SPOT

Size: 3 to 4 feet (0.9 to 1.2 m)

Range: Southern United States and northern Mexico

Habitat: Rocky terrain, desert flats, and scrublands

Diet: Lizards, eggs of birds with ground nests, rodents, and amphibians

GREAT BASIN RATTLESNAKE
(CROTALUS OREGANUS LUTOSUS)

The Great Basin rattlesnake has a thick body with a thin neck and a large, triangular head. It is light in color, often gray, brown, or yellow, with dark blotches on its back. It hibernates during winter with other Great Basin rattlesnakes in a communal burrow and emerges ready to breed. Females give birth to live young in late summer or early fall. These snakes hunt by waiting quietly for prey to amble by and striking when it appears. They coil, shake their tail, and bite when threatened.

HOW TO SPOT

Size: 30 to 35 inches (76.2 to 88.9 cm)

Range: Western United States

Habitat: Deserts and rocky hillsides

Diet: Small mammals, birds, amphibians, and insects

JACOBSON'S ORGAN

The Great Basin rattlesnake has a special receptor on the top of its mouth called Jacobson's organ. The snakes, like most rattlesnakes, detect scent with their tongues and use this organ to "smell" nearby prey or predators. Snakes that have Jacobson's organ use their nostrils to breathe and not to smell.

HOPI RATTLESNAKE
(CROTALUS VIRIDIS NUNTIUS)

The Hopi rattlesnake is named for the Hopi Native American people who are from its range in the American Southwest. This rattlesnake varies in color and matches the rocky soil where it lives. It may be pink, gray, or orange-brown with dark-brown blotches. Hopi rattlesnakes are solitary and live alone rather than in communal dens. They are diurnal and hunt for rodents and lizards in dry, rocky habitats. Each time they shed their skin, a new segment is added to their rattle.

HOW TO SPOT

Size: 24 inches (61 cm)
Range: Southwestern United States
Habitat: Deserts and semiarid areas
Diet: Small mammals and lizards

LONG-NOSED SNAKE
(RHINOCHEILUS LECONTEI)

The long-nosed snake has a V-shaped head with a long, pointed snout. It uses its long nose to help dig through soil. This snake is slender and cream-colored with black blotches. Some long-nosed snakes have red patches between the black markings. Their heads are cream with small, black spots, and their bellies are yellow or cream with no markings. Long-nosed snakes spend most of their time burrowed underground. They are secretive, nocturnal, and difficult to observe in nature. When threatened, they release a foul-smelling musk.

HOW TO SPOT

Size: 18.8 to 29.5 inches (47.8 to 74.9 cm)
Range: Southwestern United States and northern Mexico
Habitat: Deserts, grasslands, shrublands, and prairies
Diet: Lizards, amphibians, and smaller snakes

LOUISIANA PINESNAKE
(PITUOPHIS RUTHVENI)

The Louisiana pinesnake has a tan body with splotches that are brown and black. Its small head has a pointed snout. This snake is known for its loud hiss. By forcing air from its lungs to vibrate a dangling flap in its throat, this snake creates a hiss that is louder than most other snakes' hisses. Louisiana pinesnakes are constrictors that live alone. They are diurnal and hunt during the day. They spend most of their time underground in burrows hunting for gophers. Sometimes they bask above ground for warmth.

HOW TO SPOT

Size: 4 to 5 feet (1.2 to 1.5 m)
Range: Southern United States
Habitat: Pine forests
Diet: Small animals, especially pocket gophers

THE RAREST SNAKE

Populations of the Louisiana pinesnake are declining rapidly. It is now the rarest snake in the United States. Experts believe there are only about 200 left in the wild.

MANGROVE SALT MARSH SNAKE *(NERODIA CLARKII COMPRESSICAUDA)*

The mangrove salt marsh snake varies widely in color. Some are gray, brown, olive green, or tan with dark bands. Some are solid black, while others are a solid reddish orange. Some have dark stripes on the neck and some do not. The wide variations in color make this snake difficult to identify, and it is often confused with the salt marsh cottonmouth snakes that live in the same habitat. Mangrove salt marsh snakes do not have glands to help them filter the salt that enters their bodies when they eat saltwater prey.

HOW TO SPOT

Size: 15 to 30 inches (38.1 to 76.2 cm)
Range: Southeastern United States
Habitat: Saltwater marshes and mangroves
Diet: Small fish, small frogs, and fiddler crabs

MOJAVE RATTLESNAKE
(CROTALUS SCUTULATUS)

The Mojave rattlesnake varies in color so it will blend in with its surroundings. It is brown if it lives in the desert and pale green if it lives in the forest. This snake has a dark diamond pattern on its back and black-and-white bands on its tail. A light stripe runs from its eye to the corner of its mouth. Mojave rattlesnakes hibernate in winter. Mating season takes place from July to September, and females give birth to live young, often in abandoned gopher burrows. These snakes are nocturnal. They hunt at night and hide and rest in cool burrows during the day.

HOW TO SPOT

Size: 3.3 to 4.3 feet (1 to 1.3 m)

Range: Southwestern United States and central Mexico

Habitat: Scrublands, grassy plains, and mountain slopes

Diet: Rodents and lizards

AN AGGRESSIVE SNAKE

The Mojave rattlesnake has the most dangerous venom of all North American rattlesnakes. Bite victims must seek medical treatment immediately if they are to survive. This snake is aggressive toward people and other rattlesnakes and has been known to chase humans.

MOUNTAIN PATCHNOSE SNAKE
(SALVADORA GRAHAMIAE)

The mountain patchnose snake is a long, slender, tan or gray snake with two dark stripes on its back. There is a large scale on the tip of its nose that looks like a patch. Scientists believe that the snakes use this patch to help dig out lizard eggs from the soil. Mountain patchnose snakes also have large eyes with round pupils that give them excellent vision. These snakes are active and hunt during the hottest part of the day in the desert.

HOW TO SPOT

Size: 35 inches (88.9 cm)
Range: Southwestern United States and northern Mexico
Habitat: Rocky slopes and mountain woodlands
Diet: Lizards, eggs, snakes, and rodents

HIGH-ALTITUDE SNAKES

Mountain patchnose snakes live at high altitudes. They are rarely spotted below 4,500 feet (1,370 m) above sea level. They are fast-moving snakes that live in rough terrain. They can be found among trees and shrubs in rocky canyons and mountain foothills.

NIGHT SNAKE *(HYPSIGLENA TORQUATA)*

The night snake is pale gray or tan with large, black blotches on its neck and pairs of brown markings running down its back. On its flat head there are brown stripes behind its eyes and a white spot covering its mouth. As the name suggests, these snakes are active at night. Because they are hard to observe during the day, scientists know little about their social and mating habits. To protect against predators, they act like rattlesnakes. They coil their body and shake their tail, though they do not have rattles. They deliver a venomous bite with their rear fangs.

HOW TO SPOT

Size: 11.8 to 25.9 inches (30 to 65.8 cm)
Range: Western United States and Canada
Habitat: Deserts, grasslands, and suburban areas
Diet: Reptiles, amphibians, and eggs

NORTHERN COPPERHEAD
(AGKISTRODON CONTORTRIX)

The northern copperhead has a head that is reddish brown or copper colored. Its body is reddish brown with dark-brown crossbands. The bands are shaped like an hourglass, with the thin part of the hourglass marking in the center of the snake's back and the wider portions at its sides. These diurnal snakes are ambush hunters. They lie in wait for prey to appear and attack by surprise. After delivering a venomous bite, they let their prey go and follow it until it dies. Then they swallow it whole.

HOW TO SPOT

Size: 24 to 36 inches (61 to 91.4 cm)
Range: Central, eastern, and southern United States
Habitat: Wetlands, rocky hillsides, and forests
Diet: Small rodents, ground birds, lizards, large insects, and sometimes frogs and small snakes

FUN FACT

As they grow, northern copperheads lose their fangs. New fangs emerge to replace the ones they lose. Fangs are replaced about five to seven times during a snake's life.

NORTHERN RIBBON SNAKE
(THAMNOPHIS SAURITA SEPTENTRIONALIS)

The northern ribbon snake is long and slender. The top of its head is black, and its body is black or brown with three greenish stripes running down its back, like ribbons. These stripes help the snake hide in its grassy wetland surroundings. Northern ribbon snakes tend to live near water and are excellent swimmers. They hibernate in winter and emerge ready to mate in spring. Live young are born in late summer. These snakes are not considered dangerous, but when they are threatened, they release a foul-smelling musk.

HOW TO SPOT

Size: 18 to 26 inches (45.7 to 66 cm)

Range: Central and eastern United States and southern Canada

Habitat: Wetlands

Diet: Amphibians, fish, and insects

A TREE-CLIMBING SNAKE

Northern ribbon snakes spend most of their time in water, but they are also fast on the ground and very good at climbing trees. Their fast movements help them catch prey.

73

ORNATE CANTIL
(AGKISTRODON TAYLORI)

The ornate cantil, like most cantil snakes, has a thick, heavy body and long tail. Tan, orange, or yellow bands circle its black or brown body. It has small eyes with vertical pupils and a pointed snout on its broad head. Young ornate cantil snakes have a bright-green or yellow tail. They dangle and wiggle their tail tip as if it's a worm. This helps them lure prey. Adults do not have a brightly colored tail and do not lure prey this way. These snakes are nocturnal and hunt at night.

HOW TO SPOT

Size: 25 to 35 inches (63.5 to 88.9 cm)
Range: Northern Mexico
Habitat: Grasslands, tropical deciduous forests, rocky hillsides, and dry, thorny forests
Diet: Mammals, amphibians, birds, and lizards

74

PACIFIC GOPHER SNAKE
(PITUOPHIS CATENIFER CATENIFER)

The Pacific gopher snake has a dark-brown neck band and dark blotches on its tan or yellow body. Its tail is a reddish color, and its belly is yellow or cream. On its sides are smaller gray spots. Mating season takes place in spring when the snakes emerge from hibernation. Females lay clutches of about three to nine eggs during the summer. Pacific gopher snakes are easy to find on roads and trails during spring and again in the fall when young snakes hatch.

FUN FACT
Pacific gopher snakes are usually diurnal but can be active at night during extremely hot weather. They are excellent burrowers, climbers, and swimmers.

HOW TO SPOT

Size: 3 to 7 feet (0.9 to 2.1 m)
Range: West Coast of the United States
Habitat: Grasslands, woodlands, forests, farmlands, and marshes
Diet: Pocket gophers, moles, rabbits, mice, birds, bird eggs and nestlings, lizards, and insects

PLAINS BLACKHEAD SNAKE
(TANTILLA NIGRICEPS)

The plains blackhead snake is a small, slender snake. It is tan, copper, or cream with a solid black marking on the top of its head. Its belly is white with a bright-red or orange stripe down the center. Females lay small clutches in spring, and hatchlings emerge in the summer. Plains blackhead snakes are nocturnal and hunt at night. In the daytime, they hide and rest under rocks, logs, and leaves. In extremely dry conditions, these snakes will stay underground in burrows and away from the hot sun.

HOW TO SPOT

Size: 7.1 to 15 inches (18 to 38.1 cm)

Range: Midwestern and southwestern United States and northern Mexico

Habitat: Desert scrublands, arid grasslands, and woodlands

Diet: Centipedes, scorpions, and insects

PRAIRIE KINGSNAKE
(LAMPROPELTIS CALLIGASTER)

The prairie kingsnake is a large, brown or gray snake with reddish-brown blotches outlined in black. Its head is small, about the same width as its neck. Sometimes it is called the yellow-bellied kingsnake for its yellow underside. In the spring and fall when it is cooler during the day, prairie kingsnakes are diurnal. During the summer heat, they become nocturnal and hunt at night. These snakes hibernate in burrows during the winter. They emerge from hibernation earlier in the spring than most other snakes.

FUN FACT
Prairie kingsnake females do not dig shallow nests as many other snakes do. Instead, they lay their clutches under rocks or logs. The eggs stick to one another.

HOW TO SPOT

Size: 2 to 3.5 feet (0.6 to 1.1 m)

Range: Southeastern and midwestern United States

Habitat: Prairies, open woods, fields, wooded hillsides, and near farm buildings

Diet: Rodents, lizards, frogs, and occasionally snakes

REDBELLY SNAKE
(STORERIA OCCIPITOMACULATA)

The redbelly snake is brown, gray, or black. It gets its name from its bright-red belly. It is also called a fire snake because sometimes its belly is bright orange. Some redbelly snakes have a gray, brown, or orange stripe down the back. Females give birth to live young in summer. In the winter, they hibernate in shared dens, which they make in abandoned anthills, rodent burrows, and hollow tree stumps. Even when they are active, they hide under logs, rocks, and leaf piles.

HOW TO SPOT

Size: 4 to 10 inches (10.2 to 25.4 cm)

Range: Central, southeastern, and northeastern United States; southeastern Canada

Habitat: River bays, creeks, bogs, and fields and grasslands near woodlands

Diet: Slugs, insects, and earthworms

FUN FACT

When threatened, redbelly snakes can emit a foul-smelling substance to smear on their attacker. They also roll over and play dead. Their bright-red belly can startle their attacker long enough for them to escape.

RED COACHWHIP
(MASTICOPHIS FLAGELLUM PICEUS)

As its name suggests, the red coachwhip is reddish in color. It has dark blotches mixed in with banded markings on its body, and it often has dark outlines around light scales, giving it a dotted look. Its pattern fades to a solid tan or reddish color on its slender, whip-like tail. Red coachwhips have large heads and thin necks. Because they move fast, they are also called red racers. They are good climbers and dash quickly into bushes and trees when danger approaches.

HOW TO SPOT

Size: 3 to 5.5 feet (0.9 to 1.7 m)

Range: Southern United States and Mexico, including northeast Baja California peninsula

Habitat: Open pine forests, sandhill scrublands, coastal dunes, prairies, deserts, and sagebrush

Diet: Bats, birds, bird eggs, lizards, snakes, amphibians, and carrion

RED DIAMOND RATTLESNAKE
(CROTALUS RUBER)

The red diamond rattlesnake is a large snake with a long, heavy body. It can be grayish red, red, or orange. It has light markings in a diamond pattern on its back, which is how it got its name. Its tail is ringed with black and white. Young snakes are gray and grow redder as they age. Red diamond rattlesnakes are active during the day only when temperatures are mild. When the days are extremely hot, they become active at dusk and at night. Mating takes place in spring, and females give birth to live young in summer.

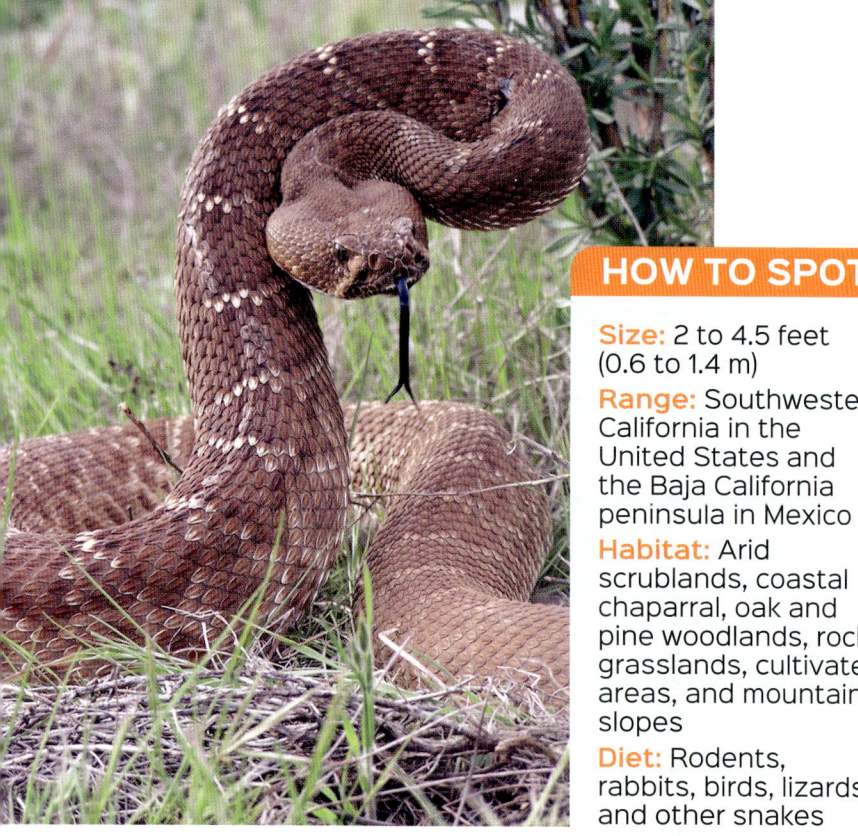

HOW TO SPOT

Size: 2 to 4.5 feet (0.6 to 1.4 m)

Range: Southwestern California in the United States and the Baja California peninsula in Mexico

Habitat: Arid scrublands, coastal chaparral, oak and pine woodlands, rocky grasslands, cultivated areas, and mountain slopes

Diet: Rodents, rabbits, birds, lizards, and other snakes

ROSY BOA
(LICHANURA TRIVIRGATA)

The rosy boa is a small snake. It gets its name from the pinkish color of its belly. The rest of its body is gray, cream, tan, yellowish, or white. Three dark stripes run along its back and sides. Rosy boas are considered calm snakes and are not likely to bite. They spend their days hiding in cool, rocky crevices. When threatened, they tuck their head and roll into a ball with only their tail sticking out. Then they squirt a foul-smelling musk.

HOW TO SPOT

Size: 1.4 to 3.7 feet (0.4 to 1.1 m)
Range: Southwestern United States and northwestern Mexico
Habitat: Deserts, rocky mountain slopes, sandy plains, arid scrublands, and brushlands
Diet: Rats, mice, young rabbits, birds, and lizards

ROUGH EARTH SNAKE
(HALDEA STRIATULA)

The rough earth snake is a slender, solid brown or gray snake with a pale ring around its neck, which fades as the snake ages. Its belly is pale yellow. Rough earth snakes can be active both day and night, but they are solitary and tend to hunt and live alone. Earthworms are their favorite food source, and these snakes are often found in damp, loose, leafy soil where earthworms thrive. Their pointed snout helps them dig into the soil for food and shelter.

FUN FACT
Rough earth snakes poop and emit a smelly musk when threatened. The foul stench helps keep trouble away.

HOW TO SPOT

Size: 7 to 10 inches (17.8 to 25.4 cm)
Range: Midwestern, south-central, and southeastern United States
Habitat: Forests and urban areas
Diet: Insects, earthworms, slugs, and snails

SHARP-TAILED SNAKE
(CONTIA TENUIS)

The sharp-tailed snake has a sharp point on the tip of its tail. It uses this point to hold down its slippery prey and to steady itself while it eats. Its small head is olive, gray, or brown, with black and sometimes orange blotches. Its back is a reddish orange, and its belly is pale with black markings. It mates in early spring, and females lay their clutch in late summer. Sometimes mothers will share nests and lay their eggs in a communal clutch. Sharp-tailed snakes spend most of their time in burrows or hiding under boards, gravel, or leaves.

HOW TO SPOT

Size: 12 to 18 inches (30.5 to 45.7 cm)
Range: Western United States and Canada
Habitat: Woodlands, forests, and grasslands
Diet: Slugs and slug eggs

SIDEWINDER *(CROTALUS CERASTES)*

The sidewinder gets its name from its unusual way of moving. It moves over the ground by throwing its body from side to side in an S-shaped curve. It moves in a diagonal line, leaving distinctive J-shaped tracks. These snakes are yellow, tan, or pale gray, with dark blotches that help them blend in with the desert sand in their surroundings. Sidewinders are also called horned rattlesnakes because they have a ridge of scales above their eyes that looks like a horn. They have a deadly, venomous bite.

HOW TO SPOT

Size: 16.9 to 31 inches (42.9 to 78.7 cm)
Range: Southwestern United States and northern Mexico
Habitat: Deserts, open flats, and rocky hillsides
Diet: Rodents and lizards

A PILE OF SNAKELETS

Sidewinders have a unique way of staying warm right after they are born. Snakelets plug the entry to their birthing den and tangle together in a big pile to share body heat. Their mother guards the den to make sure no predators get inside. After a week or so, the mother leaves, and the snakelets emerge from the den ready to live independently.

SMOOTH GREEN SNAKE
(OPHEODRYS VERNALIS)

The smooth green snake is the only bright-green snake in North America. It is solid green with no markings or blotches. This helps it blend in with its grassy environment. Its belly is white or pale yellow. The female is larger than the male. Females lay clutches from late June to early September, and hatchlings emerge quickly, between one and three weeks later. Sometimes mothers will share nests. Smooth green snakes avoid predators by blending in with the green vegetation where they live. Sometimes they bob their head to look like grass blowing in the wind.

HOW TO SPOT

Size: 14 to 20 inches (35.6 to 50.8 cm)

Range: United States, northern Mexico, and southeastern Canada

Habitat: Prairies, marshes, meadows, and pastures

Diet: Insects and spiders

FUN FACT
Shortly after a smooth green snake dies, it loses its green color and turns bright blue.

SWAMP SNAKE *(LIODYTES PYGAEA)*

The swamp snake is sometimes called the black swamp snake for its glossy black scales. Its belly is bright red or orange. It is sometimes confused with the mud snake, which also has a brightly colored belly, but mud snakes have a checkered pattern on their underside and the colors on swamp snakes are completely solid. Swamp snakes are aquatic and rarely leave the water. Females even give birth to live young in shallow water. This is unusual. Most reptiles lay eggs or give birth on land.

HOW TO SPOT

Size: 22 inches (55.9 cm)
Range: Atlantic coast of United States
Habitat: Wetlands and swamps
Diet: Fish, amphibians, and invertebrates

TEXAS BLIND SNAKE
(RENA DULCIS)

The Texas blind snake is small with a pink or brown body that is segmented like an earthworm's. In fact, it can be seen after a rainstorm and can be mistaken for an earthworm. Its eyes are two dark spots on its head, and its mouth is small. It will "stab" with its tail when it feels threatened, but there is nothing sharp to stab with and it is considered harmless to humans. This shiny, slender snake is also sometimes called a threadsnake because it is thin like a piece of thread.

HOW TO SPOT

Size: 10 to 15 inches (25.4 to 38.1 cm)

Range: Central and southwestern United States and Mexico

Habitat: Hillsides, prairies, deserts, under stones, and sometimes inside human houses

Diet: Termite and ant larvae

FUN FACT
Texas blind snakes crawl right into ant and termite nests to feed. When the insects attack them, they poop all over themselves and roll in it. The coating keeps the snakes from getting bitten, and they can feed on the insect larvae undisturbed.

TIGER RATTLESNAKE
(CROTALUS TIGRIS)

The tiger rattlesnake has a small head and a large rattle on its tail. It is light tan, gray, orange, or sometimes light purple with dark stripes, like a tiger. The tiger rattlesnake's venom is among the most dangerous of all the pit vipers. It can cause paralysis and death. Tiger rattlesnakes don't usually attack humans, so bites and injuries are rare. And when they do bite, they release only a small amount of venom. These snakes are nocturnal and hunt at night during the hot summer. They become active during the day and evening when the weather cools in the fall.

HOW TO SPOT

Size: 18 to 36 inches (45.7 to 91.4 cm)

Range: Southwestern United States and northwestern Mexico

Habitat: Desert scrublands, shrublands, woodlands, rocky slopes, mountain foothills, and rocky canyons

Diet: Lizards and small mammals, including mice

TWO-STRIPED GARTER SNAKE
(THAMNOPHIS HAMMONDII)

The two-striped garter snake can vary in appearance. Often it has a gray, brown, or green body with yellowish stripes running down its sides. Between the stripes are rows of dots. Some snakes have a checkered pattern on the sides instead of the solid stripes and do not have the dots. Two-striped garter snakes are nonvenomous, but chemicals in their saliva can cause a rash on human skin. These snakes can also be poisonous to touch or eat. The snakes absorb toxins from the toads and other animals they eat.

HOW TO SPOT

Size: 18 to 30 inches (45.7 to 76.2 cm)

Range: Western United States and Mexico

Habitat: Pools and creeks in rocky areas, woodlands, shrublands, and coniferous forests

Diet: Tadpoles, newt larvae, small frogs, and fish

WESTERN COACHWHIP
(MASTICOPHIS FLAGELLUM TESTACEUS)

The western coachwhip varies in color. It can have tan, brown, yellow, or pink smooth scales on its long, thin body. Its head is wide. This snake hunts alone and is mostly active at midmorning and late afternoon. To find prey, it pops its head up above the tall prairie grass like a periscope to see better. It also climbs trees and shrubs to get a bird's-eye view to look for prey. These nonvenomous snakes can smell with their tongue. They flick their tongue to sense odors released by other animals. This is how they can tell when prey is nearby.

HOW TO SPOT

Size: 4 to 6 feet (1.2 to 1.8 m)

Range: Great Plains and southwestern United States to south-central Mexico

Habitat: Woodlands, farmlands, desert scrublands, and prairies

Diet: Insects, amphibians, lizards, snakes, birds, and rodents

FAKE RATTLESNAKE

Western coachwhips flee to safety at the first sign of danger. If it has no place to run, the coachwhip will pretend to be a rattlesnake by coiling and shaking its tail. Sometimes predators fall for this trick and leave to avoid a "rattlesnake" strike.

WESTERN FOX SNAKE
(PANTHEROPHIS RAMSPOTTI)

The western fox snake is a long snake that looks nothing like a fox. It got its name from a scent it gives off when it feels threatened. This odor smells musky like a fox. This snake has a brown, red, or orange head. Its body is yellow with a black-and-brown splotchy checkered pattern. The western fox snake has teeth that curve toward the back of its mouth. This helps it hold and swallow large prey.

FUN FACT
The western fox snake does not have a rattle, but it can produce a rattle-like sound when it shakes its tail. Sometimes it will trick a predator into thinking it's a rattlesnake.

HOW TO SPOT

Size: 3 to 4.5 feet (0.9 to 1.4 m)
Range: Midwestern and central United States
Habitat: Farmlands, prairies, stream valleys, woodlands, and coastal dunes
Diet: Mice, voles, and birds

WESTERN HOGNOSE SNAKE
(HETERODON NASICUS)

The western hognose snake is light brown with dark-brown or gray blotches. It sometimes has orange markings on its belly. This snake is stout and stocky. It has an upturned nose like a hog snout that it uses to dig. These snakes sweep their hog-like nose from side to side, clearing away loose soil to find eggs, small animals, and insects to eat. This is also how they create burrows.

HOW TO SPOT

Size: 23.6 to 31.5 inches (59.9 to 80 cm)
Range: Southern Canada, United States, and Mexico
Habitat: Shortgrass prairies, grasslands, and rocky, semiarid regions
Diet: Amphibians, lizards, and rodents

YELLOWBELLY SEA SNAKE
(HYDROPHIS PLATURUS)

The yellowbelly sea snake has a brown back and a yellow belly. It has a flat body and a tail that looks and acts like a paddle to help it swim. Even though it spends its whole life in the water, it must have oxygen to breathe. It can take in some of the oxygen it needs through its skin while it dives and swims. Its nostrils have special valves and seals to keep out sea water. It can remain underwater for up to three hours before it needs to rise to the surface to breathe.

HOW TO SPOT

Size: 3 feet (0.9 m)
Range: Pacific Ocean off the coast of Mexico and southern California in the United States
Habitat: Salt water
Diet: Fish

SALTWATER HABITAT, FRESHWATER DIET

Although yellowbelly sea snakes live in the ocean, they drink only fresh water. Sea snakes get the fresh water they need from their food and from drinking rainwater on the ocean's surface before it mixes with the salt water.

BOLSON TORTOISE
(GOPHERUS FLAVOMARGINATUS)

The Bolson tortoise is the largest and rarest tortoise in North America. It has a high-domed, brown carapace with round, grooved scutes. It has a small, rounded head and a short tail. Its front legs have heavy claws to help with digging. Its back legs are stumpy. Males have a concave plastron. The plastron is the lower part of the tortoise's shell that covers the belly. Females have a flat plastron. Bolson tortoises burrow in the cool earth during the hottest part of the day. In winter, they hibernate in burrows. They live in colonies of up to 100 members.

HOW TO SPOT

Size: 18 inches (45.7 cm)
Range: North-central Mexico
Habitat: Arid grasslands
Diet: Grass, particularly Tobosa bunch grass

DESERT TORTOISE
(GOPHERUS AGASSIZII)

The desert tortoise has a high-domed carapace. Often the shell is greenish tan to dark brown. These tortoises have long and slender back legs. Their front legs have sharp, flattened claws for digging. With their powerful claws, they dig grooves into the earth that catch rainwater for them to drink. They also dig deep, underground burrows to hide and stay cool. This tortoise is called a keystone reptile because it is very important to its ecosystem. Other animals use the burrows they dig for shelter and protection from the desert heat.

FUN FACT
The desert tortoise is social and friendly. These tortoises like to interact with humans, and many people find them charming.

HOW TO SPOT

Size: 9 to 15 inches (22.9 to 38.1 cm)

Range: Northern Mexico and the Mojave Desert in the United States

Habitat: Sandy flats and rocky desert foothills

Diet: Grasses, herb plants, and cactus fruit and flowers

GOPHER TORTOISE
(GOPHERUS POLYPHEMUS)

The gopher tortoise has a dark-brown, gray, or black carapace. Its plastron is a lighter yellow or tan. Its front legs have large, flat claws that act like shovels for digging in the earth. Its back legs are short and stumpy. Males have two rounded scent glands under the chin to mark territory. Females do not have these scent glands. Females lay clutches from May to mid-June and do not care for their nests or their young once they emerge in early fall.

HOW TO SPOT

Size: 9 to 11 inches (22.9 to 27.9 cm)
Range: Eastern and southeastern United States
Habitat: Pine flatwoods, dry prairies, and coastal sand dunes
Diet: Berries, fruits, and carrion

TEXAS TORTOISE
(GOPHERUS BERLANDIERI)

The Texas tortoise is the smallest tortoise in North America. It has a brown, oval-shaped, flat-topped carapace with yellow or orange markings on its scutes. Its plastron is yellow. Like most tortoises, its legs are thick and sturdy. Texas tortoise legs have horned scutes that make them look a bit like a pineapple. Adults are not ready to reproduce until they are about 15 years old. Females lay small clutches of only two or three eggs. Low reproduction rates and threats to the tortoise's habitat are causing populations to decline.

FUN FACT
Texas tortoises, like all tortoises, have nerve endings in their shells. They can feel rubs, scratches, and tickles. Most tortoises like to be gently pet on their shells.

HOW TO SPOT

Size: 8.5 inches (21.6 cm)

Range: Southern Texas and northeastern Mexico

Habitat: Dry scrublands and grasslands

Diet: Plants and cactus fruit

ALLIGATOR SNAPPING TURTLE
(MACROCHELYS TEMMINCKII)

The largest of all freshwater turtles, the alligator snapping turtle is mottled in brown, black, gray, and tan. Its head and carapace feature a series of spiky scutes, which make it look like a dinosaur. Unlike most turtles, alligator turtles have eyes on the sides of their faces. They are nocturnal and scavenge and hunt for food at night. They know when prey is nearby because they can taste certain chemicals that their victims release, even underwater. These turtles spend so much time in the mud in rivers and lakes that their shells grow covered with algae.

HOW TO SPOT

Size: Up to 31 inches (78.7 cm)

Range: Central and southern United States

Habitat: Rivers, canals, streams, lakes, swamps, and wetlands

Diet: Fish, small mammals, and plants

FISHING WITH ITS TONGUE

The alligator snapping turtle has an unusual tongue that acts like a fishing lure. It will twitch its small, thin, red tongue a bit to make it look like a wriggling worm. Its unsuspecting victim will swim toward the delicious "worm" and get snapped up in the turtle's strong jaws.

ATLANTIC RIDLEY SEA TURTLE
(LEPIDOCHELYS KEMPII)

The Atlantic ridley sea turtle is the smallest sea turtle in the world. It has a gray, heart-shaped carapace and a yellow plastron. Its triangle-shaped head is gray, and it has a hooked beak. These turtles have only one claw on each front flipper. Each back flipper has two claws each. The females of this species are the only turtles that lay eggs during the day. They often return to the same spot to lay eggs where they once emerged as hatchlings.

HOW TO SPOT

Size: 23 to 28 inches (58.4 to 71.1 cm)

Range: Gulf of Mexico and the Atlantic coast

Habitat: Rivers and streams with muddy or sandy bottoms

Diet: Plants, small animals, crabs, invertebrates, mollusks, and crustaceans

FUN FACT

There were once tens of thousands of Atlantic ridley sea turtles in the Gulf of Mexico. Between the 1950s and 1980s, their population declined sharply. Scientists reported that fewer females were laying eggs. Today, their population is so low that they are considered an endangered species, at risk of going extinct.

BARBOUR'S MAP TURTLE
(GRAPTEMYS BARBOURI)

The Barbour's map turtle gets the "map" part of its name from the light-green or yellow lines on its green or black carapace. The crisscrossing lines are like those found on a map. This turtle has a yellow plastron outlined with black bars, and thin, yellow stripes on its chin. Barbour's map turtles eat snails and clams in slow-moving riverbeds, and then they move onto logs to bask in the sun.

HOW TO SPOT

Size: Males are about 4.7 inches (11.9 cm); females are up to 12.9 inches (32.8 cm)
Range: Southeastern United States
Habitat: Rivers and creeks
Diet: Mollusks, insects, and small fish

FUN FACT

Females of this species can be up to three times larger than the males. When females are born, there is more room in their shells than in males' shells so there is more room for females to grow.

BOG TURTLE
(GLYPTEMYS MUHLENBERGII)

The bog turtle is tiny. Scientists believe it is the smallest turtle in the United States. They are black or dark green with orange or yellow patches on their head. Often they can be found in mountain bogs or soggy wetlands thick with moss. Bog turtles are diurnal, but they will often burrow and rest during the hottest part of the day. When winter comes, they dive underwater and burrow under 6 to 18 inches (15.2 to 45.7 cm) of mud to hibernate.

HOW TO SPOT

Size: 4.5 inches (11.4 cm)

Range: Eastern United States

Habitat: Bogs, wetlands, and wet meadows

Diet: Plants, insects, slugs, worms, crayfish, frogs, snakes, snails, and carrion

COMMON SNAPPING TURTLE
(CHELYDRA SERPENTINA)

The common snapping turtle is a large freshwater turtle. Its carapace is black or dark green, and because it spends so much time in or near water, it is sometimes covered in moss. This turtle has a tail that is longer than its shell. The tail has bony plates that make it look like a saw. It has nostrils on top of its snout, which helps it breathe when it is almost completely underwater. Its jaw is sharp, like a beak, and its bite is dangerous. Common snapping turtles are aggressive toward humans and other animals when on land. They are more relaxed and less hostile when in the water.

HOW TO SPOT

Size: 8 to 12 inches (20.3 to 30.5 cm)

Range: Native to eastern North America from southern Canada to the Gulf of Mexico; have been introduced in western United States

Habitat: Shallow ponds, marshes, creeks, swamps, bogs, shallow lakes, and streams

Diet: Insects, spiders, worms, invertebrates, fish, frogs, small snakes, small turtles, birds, small mammals, and carrion

EASTERN BOX TURTLE
(TERRAPENE CAROLINA CAROLINA)

The eastern box turtle has a dome-shaped shell that is dark brown and black and streaked with bright yellow or orange. Its head, neck, and legs also have bright streaks of yellow or orange. It has claws rather than webs on its feet because it spends most of its time on land. Male box turtles have bright-red or orange eyes. Females' eyes are brown or yellow. Although they are diurnal, they rest in moist, cool soil in the day when it is very hot. Females lay clutches of about three to eight eggs and cover them with soil in summer, when they can be warmed by the sun.

HOW TO SPOT

Size: 5 to 6 inches (12.7 to 15.2 cm)

Range: Eastern and southeastern United States

Habitat: Forests, grassy fields, and wetlands

Diet: Plants, slugs, snails, worms, amphibians, small snakes, fruits, fungi, and carrion

A CLOSED SHELL

The eastern box turtle can close its shell completely around its body. Not all turtles can do this. Two traits make it possible: an S-shaped neck and a plastron hinge. Its curved neck can pull its entire head sideways into the shell. The hinge across its plastron allows it to close its shell tight, like a box.

FLORIDA RED-BELLIED COOTER
(PSEUDEMYS NELSONI)

The Florida red-bellied cooter, not surprisingly, has a red belly. Sometimes as the turtle ages, its red color fades. These turtles have a dark, high-domed carapace with yellow or brown markings. Yellowish stripes appear on their dark skin. These turtles spend most of their time in the water. When it's warm, they bask together on logs. They breed three to six times during the summer, and females lay clutches of 12 to 30 eggs. If females cannot find a suitable spot to nest on land, they sometimes lay their eggs in alligator nesting mounds.

HOW TO SPOT

Size: 8 to 14 inches (20.3 to 35.6 cm)

Range: Southeastern United States

Habitat: Freshwater lakes, ponds, and rivers, and sometimes brackish water

Diet: Aquatic plants

PAINTED TURTLE
(CHRYSEMYS PICTA)

The painted turtle is not hard to find in North America. It is the only turtle that can be found in every country on the entire continent. Its dark carapace is splattered with markings that are red and yellow, making it appear as though it has been dabbed with a paintbrush. Its plastron is marked with red, yellow, and orange "paint" splotches too. Painted turtles are diurnal. They forage during the day and bask on logs, rocks, and even on top of one another! They are completely inactive at night, sleeping on rocks and logs. They hibernate in winter.

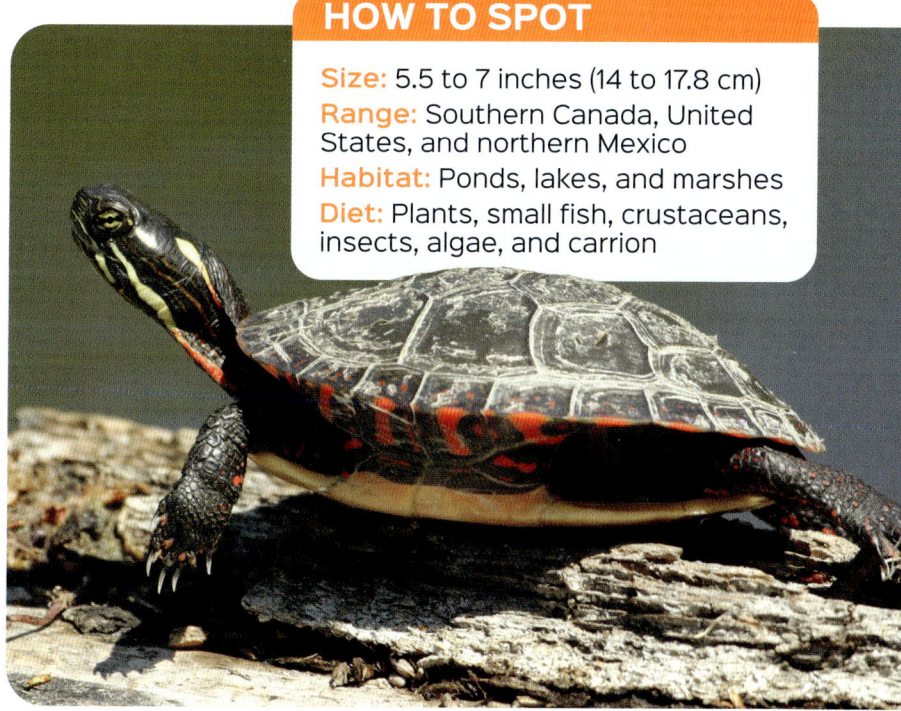

HOW TO SPOT

Size: 5.5 to 7 inches (14 to 17.8 cm)
Range: Southern Canada, United States, and northern Mexico
Habitat: Ponds, lakes, and marshes
Diet: Plants, small fish, crustaceans, insects, algae, and carrion

PLATES FOR TEETH

Painted turtles do not have teeth. Instead, they have horny plates that are rough like sandpaper. These substitute teeth help them grip their food and chew.

RED-EARED SLIDER
(TRACHEMYS SCRIPTA ELEGANS)

The red-eared slider gets part of its name from the small red stripes behind each eye where the turtle's ears should be. This turtle may have unique "ears," but it doesn't hear sound very well. Instead, it "feels" a predator's approach by sensing vibrations. When the vibrations reveal a threat, these turtles slide off logs and other basking perches and into the water quickly. This is why they are called sliders. Their olive-green carapace is shaped like an oval, and their feet have webs and claws, making them good swimmers and scavengers.

HOW TO SPOT

Size: 5 to 11 inches (12.5 to 27.9 cm)

Range: Native to central and southern United States and northern Mexico; introduced to the East and West Coasts of the United States

Habitat: Ponds, lakes, marshes, creeks, and streams

Diet: Crayfish, fish, worms, carrion, tadpoles, snails, plants, and insects

WOOD TURTLE
(GLYPTEMYS INSCULPTA)

The wood turtle does not shed the scutes of its shell, like many turtles do. The scutes grow scuffed and dull as the turtle ages. This makes the shell appear to be made of wood. Its body is brightly colored with red-and-orange markings. Its plastron is yellow with black splotches. The wood turtle is diurnal, and it basks and forages for food alone. Males are more aggressive than females and will attack other turtles, especially during mating season when they compete for the attention of females.

HOW TO SPOT

Size: 5.5 to 7.9 inches (14 to 20.1 cm)

Range: Eastern United States, the Great Lakes, and southeastern Canada

Habitat: Woodlands, wetlands, meadows, bogs, slow-moving streams, and ponds

Diet: Slugs, tadpoles, algae, plants, insects, worms, mice, eggs, and carrion

FUN FACT
Wood turtles stomp the ground to prompt worms, their primary food source, to wiggle to the surface.

GLOSSARY

autotomy
When a lizard sheds its tail to escape a threat.

brackish
A somewhat salty mixture of fresh water and ocean water.

carapace
The top of a turtle's or tortoise's shell.

carrion
A dead animal that becomes a meal for other animals.

cathemeral
The behavior in which an animal has random intervals of activity during the day or night.

clutch
A nest of eggs.

constrictor
A snake that kills its prey by coiling around and suffocating it.

dewlap
A brightly colored flap of skin on the throat of a lizard.

mangrove
A coastal tree adapted to salt water that has aboveground roots.

mimicry
When an animal has physical traits that look like a different animal.

plastron
The underside of a tortoise's or turtle's bony shell.

poisonous
Something that produces poison.

predator
An animal that kills and eats other animals.

prey
An animal that is hunted or killed.

scute
A plate that makes up the shell of a turtle or tortoise.

toxin
A substance capable of causing harm or death.

venomous
Capable of delivering a toxin by biting.

TO LEARN MORE

FURTHER READINGS

Pallotta, Jerry. *Who Would Win?: Ultimate Reptile Rumble.* Scholastic Press, 2021.

Starkey, Michael. *Reptiles for Kids*. Rockbridge Press, 2020.

Szymanski, Jennifer. *National Geographic Readers: Reptiles*. National Geographic Kids, 2022.

ONLINE RESOURCES

To learn more about reptiles, please visit **abdobooklinks.com** or scan this QR code. These links are routinely monitored and updated to provide the most current information available.

PHOTO CREDITS

ABDOBOOKS.COM

Published by Abdo Reference, a division of ABDO, PO Box 398166, Minneapolis, Minnesota 55439. Copyright © 2024 by Abdo Consulting Group, Inc. International copyrights reserved in all countries. No part of this book may be reproduced in any form without written permission from the publisher. Field Guides™ is a trademark and logo of Abdo Reference.
Printed in China

102023
012024

Editor: Leah Kaminski
Series Designer: Colleen McLaren

Library of Congress Control Number: 2023939679
Publisher's Cataloging-in-Publication Data
Names: Russo, Kristin J., author.
Title: Reptiles / by Kristin J. Russo
Description: Minneapolis, Minnesota : Abdo Reference, 2024 |
    Series: North American field guides | Includes online resources.
Identifiers: ISBN 9781098293093 (lib. bdg.) | ISBN 9798384911036
    (ebook)
Subjects: LCSH: Reptiles--Juvenile literature. | Reptiles--Behavior--
    Juvenile literature. | Reptiles--United States--Juvenile literature. |
    Encyclopedias and dictionaries--Juvenile literature.
Classification: DDC 597.9--dc23